IN BUSINESS NOW

Graphs and Charts

Renée Huggett

MACMILLAN

© Renée Huggett 1990

All rights reserved. No reproduction, copy or transmission of this publication may be made without written permission.

No paragraph of this publication may be reproduced, copied or transmitted save with written permission or in accordance with the provisions of the Copyright, Designs and Patents Act 1988 or under the terms of any licence permitting limited copying issued by the Copyright Licensing Agency, 33 – 4 Alfred Place, London WC1E 7DP.

Any person who does any unauthorised act in relation to this publication may be liable to criminal prosecution and civil claims for damages.

First published 1990

Published by
MACMILLAN EDUCATION LTD
Houndmills, Basingstoke, Hampshire RG21 2XS
and London
Companies and representatives
throughout the world

Printed in Hong Kong

British Library Cataloguing in Publication Data
Huggett, Renée
  Graphs and charts.
  1. Graphs
  I. Title II. Series
  511'.5

ISBN 0–333–51438–6

Contents

Introduction vi

Acknowledgements viii

1 Facts and Figures 1
2 Pie Charts 4
3 Vertical Bar Charts (1) 8
4 Vertical Bar Charts (2) 15
5 Multiple Bar Charts 19
6 Component Bar Charts 23
7 Single Line Graphs 27
8 Step Graphs 35
9 Multiple Line Graphs (1) 38
10 Multiple Line Graphs (2) 43
11 Histograms 47
12 Vertical Line Charts 53
13 Scatter Graphs 55
14 Straight Line Graphs 59
15 Pictograms 65
16 Cartograms 71
17 Sources of Information 76

Index 79

Introduction

The ability to create and interpret graphs and charts is an important part of practically all business examinations in schools and colleges. This student-centred book describes in simple terms:

- the basic principles of the main kinds of graphs and charts
- the information they provide and how to interpret it
- methods of construction
- the business situations in which each kind of graph or chart can be used.

The book provides the first comprehensive computer-based introduction to the subject. The instructions are non-specific so that any software can be used. Instructions are also given for drawing graphs and charts by hand.

Most graphics software is fairly easy to use as it has a range of pre-designed patterns, which can then be modified from the keyboard. Some software also provides symbols for pictograms and, even, maps.

Data can usually be imported from other programs, such as spreadsheets, or from ASCII files. The graphics can be output either to dot matrix or laser printers or to plotters.

Graphics software comes in two main forms: either as a separate program or, increasingly, in integrated business packages which include other programs such as word processors and spreadsheets.

Whatever kind of operating system is being used, a compatible program can easily be found. For example, Computer Concept's *Inter-Chart* and Minerva's *System Gamma* will both run on the BBC Micro. The same software manufacturers have produced *Inter-Chart* and *GammaPlot* respectively for Acorn's new Archimedes computer.

Digital Research's *Dr Graph* will run on the Amstrad PCW 8256 or 8512. The same software manufacturer's *GEM Graph*, which will run on IBMs and compatibles, and the BBC Master 512 (with 186 option) has many better features.

A whole range of integrated software, which includes graphics, is available ranging from Database Software's bargain-priced *Mini Office Professional* to MicroPro's *WordStar 2000*.

On the whole, the more expensive hardware and software usually provide better results and greater facilities. For example, with some software it may be difficult or impossible, to include captions to show the source of the information.

Advice on suitable software can be obtained either from your dealer or from a drop-in computer centre which have now been established by many LEAs.

This book, which will have a wide range of applications in all business education departments, will be particularly valuable for students on the following courses: GCSE Business Studies, Economics, Commerce, Computer Studies, or Information Technology; the RSA Certificate in Computer Literacy; BTEC First; TVEI extension; SCOTVEC; and YTS.

Acknowledgements

The author and publishers wish to thank the following who have kindly given permission for the use of copyright material:

The Advertising Association for material from *Advertising Association Statistics Yearbook 1987;*

The Guardian for material from *The Guardian,* 24.11.88;

The Controller of Her Majesty's Stationery Office for statistics from *Social Trends 18,* 1988 and other Crown Copyright material;

Lloyds Bank for material from *Lloyds Bank Economic Review* by Christopher Johnson, No. 115, July 1988;

London East Anglian Group and Southern Examining Group for questions from past examination papers;

Ewan MacNaughton Associates for material from the 19.10.88, *Daily Telegraph* and 26.11.88, *The Telegraph Sunday Magazine* issues;

New Statesman and Society for material from the 13.5.88, *New Society* issue;

Somerset County Council for material from Somerset County Council's Annual Report and Accounts 1987–88, published by *Somerset Express;*

Times Newspapers Ltd for material from the 9.10.88, *Sunday Times* issue.

Every effort has been made to trace all copyright holders, but if any have been inadvertently overlooked the publishers will be pleased to make the necessary arrangement at the first opportunity.

Unit 1: Facts and Figures

Facts and figures can be presented in many ways. Let's say you were investigating the effects of transport on the environment. Information about lead in exhaust fumes of vehicles could be presented either in a table (Figure 1.1) or in a bar chart (Figure 1.2). Which is the more eye-catching?

Emissions of lead from vehicles

	thousand tonnes
1981	6.7
1982	6.8
1983	6.9
1984	7.2
1985	6.5
1986	2.9

Source: Social Trends 18

Figure 1.1 *Information can be presented in a table*

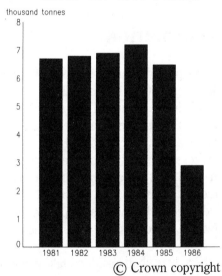

Figure 1.2 *The same information presented in a bar chart (source: Social Trends 18)*

Advantages

The chart has many advantages compared with the table.

- It attracts attention, because it is more striking.
- It shows facts, such as the dramatic fall in 1986, more clearly.

- It is easy to understand.
- It saves the reader's time because it provides the information at a glance.

That is why graphics, or visual means of presenting information, are now used so much by business, the government and newspapers and magazines. It is the modern means of communication. As the Chinese say, one picture is worth a thousand words.

Disadvantages

Charts and graphs, however, do have some disadvantages compared with tables. The information is not always so precise, or accurate.

Look at Figure 1.2 again. It is not easy to tell exactly how many thousand tonnes of lead were given off by petrol-driven vehicles each year. The chart simply shows that the amount rose from nearly 7 thousand tonnes to just over 7 thousand tonnes a year between 1981 and 1985. It doesn't give the actual figures as does the table. How could this problem be overcome in a bar chart?

Limited information

Charts (and tables) provide only a limited amount of information. They state *what* has happened; they do not tell the reader *why* it has happened, or what may happen in the future. They show one set of facts, but do not tell the whole story. For that reason, great care is needed in interpreting, or explaining the meaning, of charts and graphs.

Figure 1.2 above shows only that the amount of lead remained fairly steady between 1981 and 1985 and then fell suddenly in 1986. Without further information, it is impossible to say what is likely to happen in the future. We would need to know what happened in the years after 1986 to see if the downwards trend or pattern continued, or whether it was just a chance event.

However, as there was such a sudden fall, it seems likely that there was a special reason. In fact, the fall occurred because a new law came into effect at the end of 1985. This reduced the permitted amount of lead in petrol from 0.4 gram to 0.14 gram per litre.

Using this piece of information we can now *measure* the effect of the law in reducing the amount of lead given off in exhaust fumes.

It is important to make sure that you have all the relevant information before you try to explain the meanings of charts and graphs. Otherwise, you may jump to hasty conclusions which are quite likely to be wrong.

Facts and Figures 3

Simple to use

An ability to use graphs and charts is now essential in business. It is very easy to create them with a computer and graphics software. Most software is very simple to use, as it draws the graph or chart and does most of the calculations for you. After a little practice, you should soon be able to create professional-looking charts and graphs, like the illustrations in this book. Most of them were created with a software package called Dr Graph.

However, you will still need to know how to interpret graphs and charts and when they should be used. These matters are also covered in detail in the following units.

Activities

1

Find three different kinds of graphs or charts in newspapers and magazines. State what they are called; or, if you do not know already, look through this book until you find the answers.

2

Find a chart or a graph in a newspaper or a magazine which seems to give inaccurate information. State why it is inaccurate and how you would change the chart or graph.

3

What was the percentage decrease in lead emissions between 1985 and 1986? What other information would you need, before you could calculate the real extent of the fall?

Unit 2 Pie Charts

A pie chart, like Figure 2.1 below, is drawn as a circle which is divided into a number of segments or slices. It shows the size of each slice as a part of the whole.

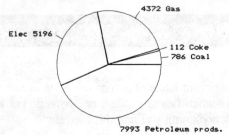

Figure 2.1
A pie chart shows the size of each slice (source: Department of Energy*)*

Note that charts (and graphs) should have a title and state the source of the information. If your software does not write captions, you will have to type or print the source of the information below the chart.

If you are taking an examination, it is particularly important to include both a title and the source of the information. In business documents, advertisements and newspaper reports, however, the title or the source is sometimes omitted, or left out. They are also often left out in books, when the chart is being used to illustrate some principle.

Main uses

Pie charts are most useful when the whole divides into five, or fewer, slices of different sizes, as in Figure 2.1 above.

A pie chart could also be used to show:

(a) different kinds of households, such as persons living alone, married couples, married couples with children, single-parent families.
(b) tenure of homes, such as owned, mortgaged, rented, housing association, service tenancy where the home is provided as part of a job.

(c) firm's sales by region, such as South-east, North-west, North-east, Midlands, South-west,
(d) current liabilities of a firm, such as overdraft, loans, creditors, taxation.

Software makes it very easy to have a large number of slices. For example, in (a) above, you could have 10 slices to show bachelors, spinsters, widows, widowers, married couples without children, married couples with one child, married couples with 2 children, married couples with more than 2 children, one-parent families with a male head of household and one-parent families with a female head of household. However, the reader would find it difficult to take in the information at a glance. In fact, that is one of the dangers of software: it is so easy to present a mass of information that you may be tempted to include too much.

You should never clutter up any chart or graph with so much information that the reader becomes confused.

Look at Figure 2.2 below. There are far too many slices and some of them are so small that it is difficult to judge their relative sizes. If the percentage figures were not included, it would be almost impossible to see the difference.

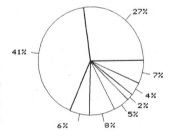

Figure 2.2
A pie chart should never contain too many slices

Percentages

If you key in the actual numbers for a pie chart, as in Figure 2.1, your software will automatically work out the correct size of each slice. This saves you a lot of time and trouble. However, if you want to use percentages instead, you may have to work out the figures for yourself.

To convert the data in Figure 2.1 to percentages, you would first have to work out the total amount that consumers spent on energy by adding up all the figures. Then use the following formula to calculate the percentage share for each item:

$$\frac{\text{Individual amount}}{\text{Total amount}} \times 100$$

For example, the percentage share of gas was:

$$\frac{\pounds 4372 \text{ million}}{\pounds 18{,}459 \text{ million}} \times 100 = 23.7\%$$

Now work out the percentages for the other items in Figure 2.1.

Comparisons

Pie charts should not be used to compare the same items for different years, such as the amounts spent on energy in 1986 and 1987, or any later year. The total amounts spent in each year vary, so to give a true picture of the facts you would have to use circles of different sizes. Even if you did, it would still be difficult to see the differences in the relative size of slices for each year.

If you want to make comparisons of the same items over a period of time, a component bar chart is far more attractive (see Unit 6).

On the whole, therefore, you should use a pie chart only when you want to present a limited amount of information for one point in time. As long as there are only four of five slices, the pie chart can have quite a dramatic impact.

Hand-drawn charts

Pie charts are fairly easy to draw, but you will have to do some calculations first. First, you need to work out the angle for each segment or slice. You do this by using a formula similar to the one for percentages above, but using 360 instead of 100, because there are 360° in a circle. The formula is:

$$\frac{\text{Individual amount}}{\text{Total amount}} \times 360$$

So, the angle for gas in Figure 2.1 would be:

$$\frac{\pounds 4372 \text{ million}}{\pounds 18{,}459 \text{ million}} \times 360 = 85°$$

Work out the angles for the other slices in Figure 2.1, and round them off to the nearest whole number so that the total is 360°.

Use a compass and a protractor to draw the chart. Add a title, the amounts for each slice, and the source of the information.

You could colour each slice differently, if you wished.

Activities

1

Create a pie chart for consumer spending on energy for 1984, using the following Department of Energy figures:

	£ million
Coal	659
Coke	106
Gas	3,761
Electricity	4,551
Petroleum products	8,025

2

The current liabilities of a public limited company (plc) in the previous year were:

	£000
Overdraft and loans	8,687
Creditors	37,183
Tax	4,991

Create a pie chart to illustrate the current liabilities.

3

Look at the pie charts showing the income and spending of Somerset County Council for 1987–88 (Figure 2.3) and answer the following questions.
(a) Which was the biggest source of income?
(b) What was the biggest item of expenditure?
(c) How clearly, in your view, is the information presented in the pie charts?

Analysis of Gross Expenditure

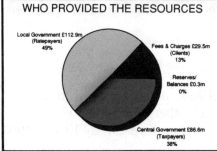
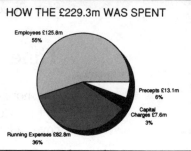

Figure 2.3
(Source of charts: Somerset Express – for Somerset County Council)

Unit 3 Vertical Bar Charts (1)

Vertical bar charts provide a very clear and dramatic way of presenting information. That is why they are so widely used in business. They are ideal for showing changes in the same item over a period of years, such as the rise in the percentage of homes with a video recorder.

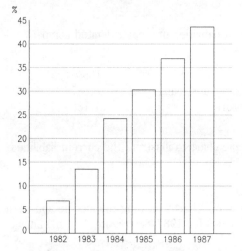

Figure 3.1
Bar charts are ideal for illustrating changes over time (source: New Society, *13 May 1988)*

Charts of this kind can be used for many similar topics, such as changes over the years in:

- the number of unemployed
- the profits of a firm
- the amount of credit
- the number of word processors used in offices
- the number of new factories opened in an area.

State three other topics which could be illustrated with a vertical bar chart.

Time

The horizontal axis, or line, is used to show time. Consecutive years, which follow each other, are used in Figure 3.1 above. If you wanted to cover a larger number of years, say 1960 to 1990, you would have to use

1960, 1965, 1970 etc. on the horizontal scale. Without these bigger intervals, you would not be able to get all the years on the scale.

Software will choose suitable intervals for you or you can select them yourself. What intervals would you use if you wanted to cover the years from 1961 to 1989? Other periods of time can be used instead of years. When the rate of change is very fast, shorter periods can be used instead: quarters of a year, months – or even days, during periods of crisis.

For example, during the stock market crash of October 1987, when the value of shares dropped daily by billions of £s, newspapers used bar charts to show the falls day by day. State two other situations in which days might be used on the horizontal axis.

If you use days on the horizontal axis, software usually abbreviates, or shortens, the name of the days so that they do not take up too much room on the horizontal scale. So, Monday is abbreviated to Mon. Software sometimes provides a choice of one-letter abbreviations, such as M for Monday, so that more days can be fitted on to the horizontal scale. The names of months are shortened in a similar way.

Vertical axis

The scale on the vertical axis, or line, consists of a series of marks at regular intervals – like a ruler. It helps the reader to measure the length of each bar, so that they can be compared more easily (note that the *width* of each bar must always be exactly the same). It also provides a rough guide to the value of each bar; for example, in Figure 3.1, 1986 is under 40% and 1987 is over 40%. Percentages are used in Figure 3.2; but actual numbers could be used instead. In some cases, such as a firm's profits, a £ scale would be used on the vertical axis.

Better design

Software makes it very easy to improve the design of a chart so that it looks even more striking. By tapping a few keys, you can do in seconds what would take you hours by hand.

The main aims of design are:

- to make it easier for the reader to understand the chart at a glance
- to attract attention
- to produce a pleasing appearance.

Look back at Figure 3.1. Now, let's see some of the ways in which the design and appearance can be improved.

Guide lines

The addition of dotted guide lines, as in Figure 3.2 below, makes it easier for the reader to judge the height of each bar. The lines are created by extending the scale on the vertical axis right across the chart. Most software gives a choice of styles for guide lines, or grid lines, such as

— — — — — — — — — —

....

In this case, a dotted line is used. It is just as effective as the other grid line in guiding the reader's eye to the scale on the vertical axis, and doesn't divert attention from the information.

If you wanted to provide the reader with more accurate information, the bars could be labelled with the actual percentages, as in Figure 3.3 below.

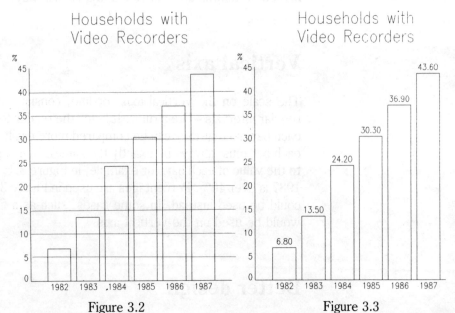

Figure 3.2
Grid lines help to guide the reader's eye

Figure 3.3
Do not use grid lines if you include numbers

Figure 3.2   Figure 3.3

With some software, you do not have to key in the figures again. The computer will use the figures you entered when you created the chart. When you tap in the command, the numbers will appear at the top of each bar as if by magic.

For greater accuracy, you could also state in a note below the chart whether the percentages are an average for the year or if they refer to some particular month.

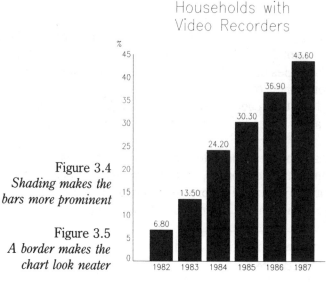
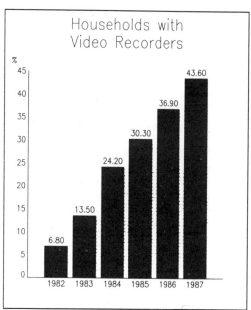

Figure 3.4
Shading makes the bars more prominent

Figure 3.5
A border makes the chart look neater

Figure 3.4

Figure 3.5

Shading

Shading the bars, as in Figure 3.4 above, is another great improvement. The shading makes the bars stand out more and attracts attention.

You usually have a choice of four or more different kinds of fill patterns, or kinds of shading. They range from a black, solid fill, as in Figure 3.4 above, to a much lighter shading. Sometimes, the software provides various kinds of cross hatching instead. What you get depends on the computer operating system, the software and the printer or plotter you are using.

In this case, the same shading should be used for all the bars, as only one kind of fact is being presented.

If you are using a computer with a colour display monitor, colour could be used instead of shading to produce coloured on-screen graphics.

Putting a border around the chart is another big improvement that can easily be made with software, as in Figure 3.5 above. It helps to tie the whole chart together, and makes it neater.

You can usually alter the width of the border. The thicker the border, the more the chart will stand out on the page.

If you wished, you could have a border around the axes instead, so that the title stands outside the border.

Titles

The title should be bigger than any other text in the chart. Software allows you to change the size of the title, or any other printing in the chart, such as the numbers on the scales or the descriptions of the axes. It also provides a choice of different fonts, or kinds, of type.

The title should tell the reader what the chart is about. It should be as brief as possible and written in a way which will appeal to the intended readers.

With such a heavy border, the title now seems rather insignificant or weak. You could use a thinner border, or you could use a stronger type for the title as in Figure 3.6 below. This provides a better balance of black and white on the page. It also helps to draw attention to the title.

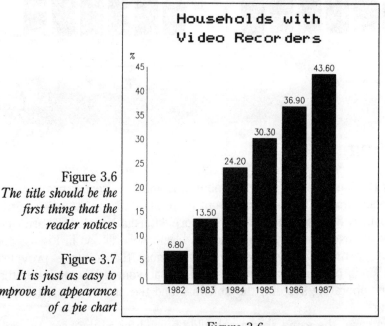

Figure 3.6
The title should be the first thing that the reader notices

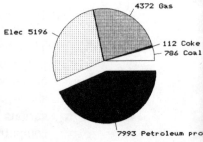

Figure 3.7
It is just as easy to improve the appearance of a pie chart

Figure 3.6   Figure 3.7

Similar kinds of improvements can also be made to pie charts, as in Figure 3.7 above. Slices can be shaded to distinguish them from each other. The biggest slice – Petroleum Products – can be exploded, or taken out of the pie, to attract the reader's attention. The slices can be put in either ascending or descending order of size. Text in the chart can be moved to other places.

Look back at Figure 2.1. Describe the changes that have been made in the improved version in Figure 3.7 above. What further improvements would you make?

Hand-drawn charts

It is possible to draw quite reasonable bar charts by hand. All you need is a ruler, a pen or pencil – and a lot of care!

The bars should be of exactly the same width. They should not be too long and narrow, or too short and fat.

A space should be left between each bar. The space should be up to half the width of a bar. All the spaces must be of the same width.

To improve the design, you could draw a border around the whole of a chart. If you are careful, you could colour or shade the bars. With simple bar charts which show only one kind of data, the same colour or shading should be used for each bar.

Activities

1

(a) For which kinds of publications would the title in Figure 3.6 above be suitable?

(b) If the present title were used as a smaller sub-title, write a main title which could be used in an advertisement trying to increase the sales of video recorders.

2

The following figures show the percentage of households with central heating for five consecutive years.

    1983   47%
    1984   55%
    1985   55%
    1986   56%
    1987   60%
(Source: *New Society*, 13 May 1988)

(a) In which year was there no increase?
(b) Which year had the biggest percentage rise?
Create a vertical bar chart with a main title suitable for a Government publication.

3

Create a bar chart based on the following table which forecasts the number of young people from 16–19 in the working population from 1991 to 1995. Provide a main title which would be suitable for a young people's magazine.

	Thousands
1991	2,204
1992	2,096
1993	2,006
1994	1,961
1995	1,968

(Source: *Employment Gazette*, May 1988)

(a) Look at the chart and explain in words what the chart shows about youth employment in the future.
(b) What are the effects of the change likely to be on:
   (i) wages of young people?
   (ii) employers?

Unit 4 Vertical Bar Charts (2)

A vertical bar chart was used in the last unit to show changes *over a period of time*. Bar charts can also be used for comparing the same item in various countries *at one point in time*, as in Figure 4.1 below.

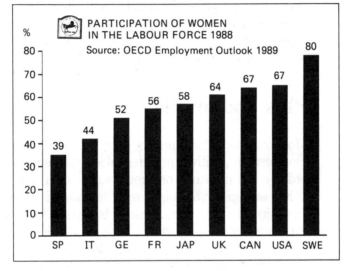

Figure 4.1
A vertical bar chart can be used to make comparisons between countries (source of chart: Lloyds Bank Economic Bulletin, *July 1988)*

There are several points to note about Figure 4.1 above:

(a) The scale on the vertical axis is marked in percentages.
(b) Each bar refers to a country (or geographical location). This is indicated on the horizontal axis.
(c) The bars are usually drawn in ascending order, with the smallest first.
(d) To provide more accurate information, the actual percentage is given at the top of each bar.
(e) To save space, the names of the countries are abbreviated, or shortened.

What are the full names of the countries in the chart? How would you abbreviate Poland, Belgium, Paraguay, the Netherlands?

You could create similar bar charts comparing the number of working days lost by strikes, the percentage of homes with microwave ovens, the pupil/teacher ratio (or the number of pupils for each teacher) in different countries. Give two other examples.

Comparisons

Charts of this kind can also be used to make comparisons between other geographical locations, such as areas of a country. So, you could create a chart showing the percentage of consumers who own video recorders in various regions of the country. The percentage would be on the vertical axis, and the regions of the country on the horizontal axis.

They can also be used for comparing different business sectors, such as agriculture, mining, manufacturing etc., in relation to profits or to the numbers of people employed. In these cases, the scale on the vertical axis would be £s if you were dealing with profits, or numbers if you were dealing with the size of the workforce. Give three different examples of when such charts could be used.

Plus and minus

Most bar charts, whether they are showing comparisons or changes over time, deal in positive, or plus, values. Sometimes we need to show both positive and negative, or minus, values over a period of time.

For example, you might want to show the profit and loss made by a small firm during its first year of business. The financial results could be illustrated by a bar chart like Figure 4.2 below.

Note that the vertical axis and scale is extended, or continued, below zero. Once again, most software makes it very easy for you to create charts of this kind. Simply key in the numbers, marking the negative values with a minus sign (−), and the software will draw the chart for you!

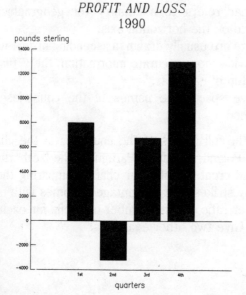

Figure 4.2
Bar charts can show both plus and minus values

Vertical Bar Charts (2)

Deviation bar charts are very useful for illustrating financial matters, where there are often both plus and minus values. They can be used to show:
- variations in the country's balance of payments over the years
- a business's cash flow forecasts
- a firm's profit and loss over months, quarters or years
- the rise and fall in the government's public borrowing requirement.

State two other examples of when a deviation bar chart could be used.

Hand-drawn charts

Deviation bar charts can be drawn by hand. Be careful to use the same scale on the vertical axis for both positive and negative values. As only one kind of data is being used, the same colour or shading should be used for each bar.

Activities

1

Use the following figures to create a vertical bar chart showing the amounts that were spent on advertising in the Press between 1980 and 1986.

Advertising Expenditure
The Press

Year	£m	% of total
1980	1,684	65.9
1981	1,816	64.4
1982	1,986	63.5
1983	2,236	62.5
1984	2,558	63.0
1985	2,801	63.1
1986	3,136	61.3

Source: *Advertising Association Statistics Yearbook 1987*

Answer the following questions:

(a) In which year was most money spent on advertising in the Press?
(b) How do you account for the fact that the amount spent on Press advertising has risen, while its share as a percentage of the total has been falling?
(c) What, in your view, is the likely reason for the decline in the percentage spent on Press advertising?

18 Graphs and Charts

2

Create a deviation bar chart showing the variations in an imaginary country's balance of payments, using the following figures:

	$
1988	10,210,000,000
1989	−2,130,000,000
1990	7,999,000,000

3

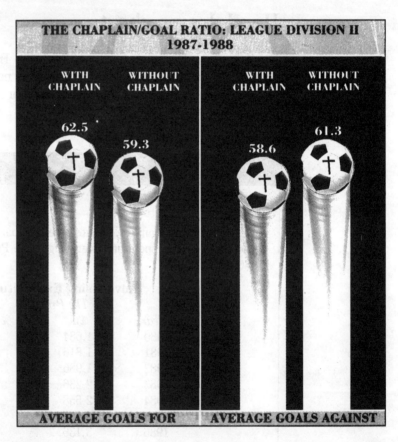

Figure 4.3
(*Source of chart:* Daily Telegraph Weekend Magazine, *26 November 1988*)

What are your views of (a) the method of presentation and (b) the information?

Unit 5 Multiple Bar Charts

Multiple bar charts are extremely useful if you want to present a large amount of information in a small amount of space. Instead of having only one bar at each division on the horizontal axis, there are two (or more) bars, as in Figure 5.1 below.

It is essential that the reader can see at a glance what each bar represents, or stands for. One way of doing this is to label each bar. For example, if the bars represented men and women you could label them 'Men' and 'Women', or with the abbreviations 'M' and 'W' as in Figure 5.1.

It is much better – and much easier with most software – to use contrasting shading for each bar as in Figure 5.2 below. The different shading makes the chart look far more interesting.

Like the key on a map, the legend at the top of Figure 5.2, explains what the shading means. If the legend gets in the way of one of the bars, it can usually be moved to another part of the chart with a couple of keystrokes.

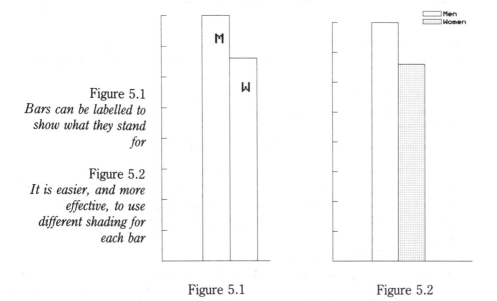

Figure 5.1
Bars can be labelled to show what they stand for

Figure 5.2
It is easier, and more effective, to use different shading for each bar

Figure 5.1   Figure 5.2

In choosing the shading for the bars, you should aim for contrast: a light shade, followed by a medium shade, followed by a dark shade. This improves the design and makes it much easier for the reader to see which bar is which.

Number of bars

You must be careful never to have too many bars for each year. For example, if you were illustrating the kinds of homes in which people live, you could have bars for every kind of home, such as detached, semi-detached and terraced houses, flats, caravans, mobile homes, rooms, even palaces! If you had so many bars, the reader would simply become confused.

If there are a large number of items like this, there is a simple way around the problem. Choose the main kinds of homes in which most people live, that is, houses and flats. Then include all the other items in one class called 'others'. In that way, you will have only three bars for each year – houses, flats and other kinds of homes.

Some software allows you to have five or more bars for each year, but it is best never to use more than three.

Advantages

Multiple bar charts are very useful for comparing the same items in different years.

- The data does not necessarily have to form part of a whole. You can show the relationships between any two, or more, parts of a set of data.
- It is much easier to see the difference between the length of bars than it is to judge the relative size of slices in a pie chart.
- You can produce a well-designed and attractive-looking chart even if the bars are of very different lengths as in Figure 5.3 below.

Note that the bars are quite narrow so that they do not take up much more space than a single bar. Your software will adjust, or change, the width of the bars automatically. The more bars you use, the narrower they will be. Most software also allows you to change the width of the bars yourself, if you wish.

Figure 5.3 provides a great deal of information. Answer the following questions by using the chart and your own knowledge.

(a) Which is the most popular part of the world with British visitors?
(b) Which is the second most popular? Suggest where some of these places might be.
(c) Approximately how many visits abroad did British people make in 1983 and 1987? What was the percentage increase?
(d) What happened to the number of visits to North America in 1984 and 1985 compared with the previous and following years? Suggest reasons for the change.

Activities

1

Look at the latest volume of the *Annual Abstract of Statistics* in your main public library. Obtain some contrasting figures about men and women for the last five years in relation to any topic such as employment, wealth, divorce, education etc. Create a multiple bar chart based on the information you have obtained.

2

You have found out the following information about a firm:

Last year
Turnover   £386,782
Profit   £89,023

This year
Turnover   £593,341
Profit   £134,660

Construct a multiple bar chart based on this information. In which year was the profit higher:

(a) in pounds (£)
(b) as a percentage of turnover.

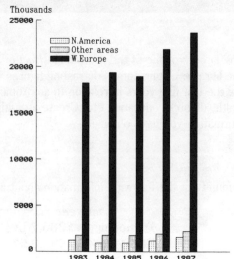

Figure 5.3
Multiple bar charts provide a large amount of information in a small space (source: Employment Gazette, September 1988 (Extracts))

© Crown copyright

(e) How many visits on average did each British person make abroad 1987? Describe the effects on Britain's balance of payments.

Main uses

Multiple bar charts (which are sometimes called compound bar charts) are very useful for showing changes in two or three items over periods of time. They could be used to show changes in:

- unemployment among men, women and young people
- the kinds of books borrowed from public libraries – fiction and non-fiction
- sources of credit – bank, building societies and hire purchase
- purchases of new and second-hand cars.

Give two other examples.

Hand-drawn charts

Multiple bar charts can be drawn by hand, though you must be careful to make all bars exactly the same width.

You will need to use a legend, or label the bars, to explain what each one represents. Colour, or shading, can be used to distinguish the bars.

Unit 6 Component Bar Charts

Another way of presenting data about different items is to use a component bar chart, like Figure 6.1 below. Instead of using two or three separate bars as in a multiple bar chart, all the data for each year is contained in a single bar. Each component, or part, of the bar is shaded differently. A legend is again used to provide a key to the meaning. You could label each component instead: but that would take far more time with most software.

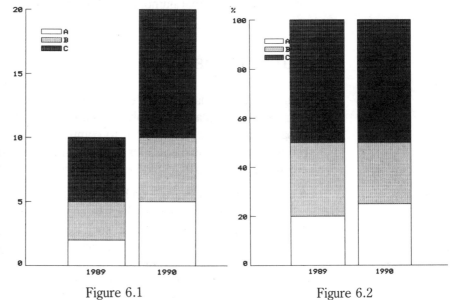

Figure 6.1
A component bar chart shows numbers

Figure 6.2
The same data presented in a percentage bar chart

Figure 6.1   Figure 6.2

Disadvantages

These charts have one big disadvantage. It is not always easy to compare the size of the components, or parts, for each year.

Look at Figure 6.1 again. It is easy to see the difference in size between Component A in 1989 and 1990 because they both start on the same baseline of zero, or 0.

Now look at Component B. It is obvious that it increased in 1990; but it is more difficult to say by how much, because the components start at different points on the vertical scale.

Work out from Figure 6.1 the size of each component for both years.

Percentage bar charts

One way of getting round this problem is to use a percentage bar chart instead, as in Figure 6.2 above. In these charts, the length of the bars is the same, because they represent the whole amount, or 100%. Percentages are used, instead of actual numbers, for the area of each component.

In Figure 6.1, it seems that Component C has increased by the greatest amount – and so it has in actual numbers. From Figure 6.2, however, it can be seen straight away that Component C's share of the whole has remained the same between 1989 and 1990. Work out the percentages for each component for both years.

The main problem with percentage bar charts is that they do not show the increase in actual numbers in the two years.

Advantages

Component bar charts do have one big advantage: they are more effective than any other chart or graph for showing changes in a large number of separate items. So, if you want to compare a large number of items over the years, you should consider using a compound bar chart. For example, a compound bar chart could be used to show:

- Consumer expenditure on housing, food, drink, tobacco, fuel, clothing and footwear, transport, durable consumer goods etc.
- Spending by local councils on such items as education, social services, roads, police, fire services, libraries, waste disposal, planning etc.
- National output of different business sectors, such as agriculture, forestry and fishing, mining and quarrying, manufacturing, construction, distribution, financial services etc.
- Trade with various areas of the world, including EC countries, other European countries, North America, developing countries etc.

Even so, you shouldn't include too many components. If you do, the reader may find it difficult to understand the chart. Furthermore, your software may run out of different kinds of shading, and you will have to label each of the components instead, which will take you much longer!

Contrasts

These charts are most effective when they show big contrasts between the years, as in Figure 6.3 below. This chart illustrates the changes in the use of road and rail transport over a 25-year period.

(a) What has become the most popular form of transport during this period?

Component Bar Charts

(b) Has the distance travelled by cyclists increased or decreased?

(c) Why, in your view, has the amount of travel by rail remained about the same?

Note that the smallest item – bicycles – has the darkest shading. If it were left unshaded, or had only light shading, the reader might not notice it as much.

Horizontal bar charts

Any bar chart can be drawn with horizontal, instead of vertical bars. Horizontal bars are often used when the chart contains text, as in Figure 6.4 below. As we read the text from left to right, it is more natural to have the bars going in the same direction, too.

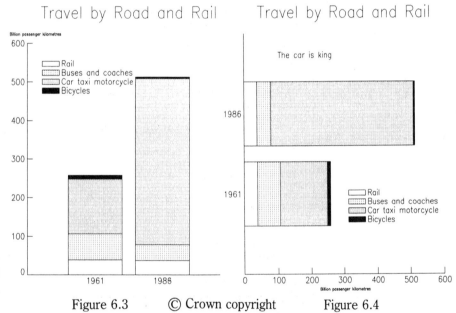

Figure 6.3
Component bar charts stack one item above another (source: Social Trends 18, *1988)*

Figure 6.4
Horizontal bars and a slogan

Software makes it very simple to swing the bars around from vertical to horizontal. Figure 6.3 was changed into Figure 6.4 with a single tap on a key which changed V to H!

With most software, you can change the size of the bars. The bars in Figure 6.4 are rather fat. How much thinner would you make them?

Use sparingly

Component bar charts have their place, but you should use them only when you want to include a large number of items in a chart. They are more difficult to construct and to read than many other charts and graphs.

26 Graphs and Charts

If you do not have the use of a computer and graphics software, you would be wise not to attempt them at all!

Activities

1

The data for Figure 6.4 is:

	billion passenger kilometres	
	1961	*1986*
Rail	39	37
Buses and coaches	67	41
Cars, taxis, motorcycles	142	430
Bicycles	10	4

Create a component bar chart with horizontal bars, like the one in Figure 6.4, using narrower bars.

2

You have obtained the following figures about the money spent by a council over two years:

	1990	*1991*
	£m	*£m*
Education	223	246.9
Social Services	42.8	43.6
Roads	21.3	29.8
Police	11.9	12.8
Fire	7.4	7.3
Waste disposal	2.1	2.0
Other services	0.5	0.2
Central Services	17.4	12.1

Construct a component bar chart with vertical bars. Which service has increased most in cost:

(a) in pounds (£)
(b) as a percentage?

Which service has shown the biggest *percentage* fall in cost?

Unit 7 Single Line Graphs

The same data can sometimes be presented in either a chart or a graph. Each will have a different impact on the reader.

For example, a bar chart (shown in Figure 7.1 below) was used in Unit 4 to provide information about video recorders. Because the bars are separate rectangles, they focus the reader's attention on the percentages, which can be read off on the vertical axis.

The same facts can be presented in a line graph, as in Figure 7.2 below. Because there is only a line, or curve, the reader's attention is directed more towards the upwards trend in the number of homes with video recorders. (In graphs, the line is usually called a curve, whether it is straight or curved.)

In business, graphs are usually used to show trends, or the way things are going, over a period of time.

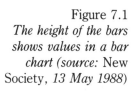

Figure 7.1
The height of the bars shows values in a bar chart (source: New Society, *13 May 1988)*

Figure 7.2
Values are read off the curve in a graph (source: New Society, *13 May 1988)*

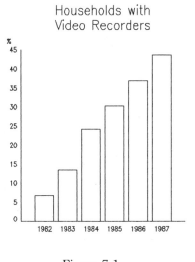

Figure 7.1

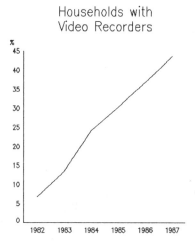

Figure 7.2

Note that the scales on the vertical and horizontal axes, or lines, are exactly the same in both the chart and the graph. The only difference between them is that the graph uses a curve, instead of vertical bars.

Design

Software makes it very easy to improve the appearance and design of the line graph, just as was done with bar charts. The addition of horizontal grid lines in Figure 7.3 makes it somewhat easier to read the percentages on the vertical scale.

The curve has also been thickened to make it bolder and easier to see. This also helps to iron out any irregularity, or unevenness, in the curve so that it looks straighter.

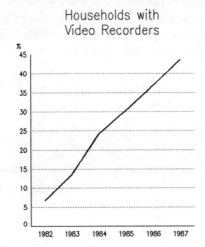

Figure 7.3
A thicker curve is easier to see

Figure 7.4 below shows other improvements that can be made. The addition of vertical grid lines make it easier to read off the years on the horizontal scale. A border round the axes helps to make the graph look neater. (A border round the whole graph could have been used instead.)

The border, however, makes the previous title appear somewhat weak in comparison, so a bolder type is used for the title to make it stand out more.

Interpretation

Great care is needed in interpreting line graphs. You cannot read off values at all points on the graph, but only at the divisions on the horizontal scale.

Look at Figure 7.4 below. You can see that the percentage of homes with video recorders in 1982 was roughly 7%. What were the percentages, roughly, in 1984 and 1987? These pairs of figures – 1982 and 7 etc. – are known as the coordinates. They are like the coordinates on a map.

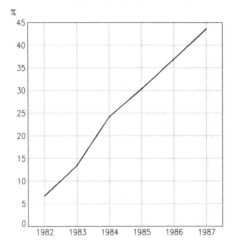

Figure 7.4
Grid lines make it easier to read off values on the axes

Although it is safe to read off these values, you cannot use the graph to read off values at intermediate points on the scales. For example, you could not find out the percentage figures for July of each year.

The graph only provides data for each year (which may be an average or one particular point in the year). There is no way of telling what happened in the months between. For all you know, there may have been a steady sale of video recorders throughout the year, or they may all have been bought in the months before Christmas! (There are some straight-line graphs, where it is safe to read off intermediate values – see Unit 14.)

In a similar way, you must be careful not to assume that the trend shown by the curve will continue at its present rate. For example, in Figure 7.4 above, you cannot assume that the curve will continue to climb upwards. It may go on rising until almost every home has a video recorder, or it may flatten out, if sales have already reached their peak.

Variables

A line graph is always used to show the relationship between two variables – or things which can be measured in numbers. Time, for example, can be measured as a number of years, months, weeks, days etc. The total of video recorders in a country can be measured as a number, or as a percentage of the homes that have one. Population can be measured in numbers; sales can be measured in numbers of £s or as an actual number of goods that have been sold.

There are other things that cannot be measured in numbers. For example, the quality of service in a shop cannot be measured, so it is not a variable. On the other hand, the complaints that a shop receives about the service it provides can be measured in numbers, so it is a variable.

State six other variables which might be used in a business graph.

Two kinds of variable

There are two kinds of variables: independent and dependent. An independent variable is one which is not affected by the other variable. For example, time is an independent variable because it changes at the same rate whether we want it to or not. One second is followed by another second, one minute by another minute, regardless of whether we stop the clock or not.

Distance is another independent variable. Two miles is twice as long as one mile. The distance is measured and fixed, and cannot be affected by any other variable. In graphs, the independent variable should always be shown on the horizontal axis. If time is one of the variables, it is always shown on that axis.

The dependent variable is always shown on the vertical axis, for example, the value of sales, the number of video recorders etc.

So if you were creating a graph to show the costs of running a lorry per mile, the distance (or independent variable) would be on the horizontal axis and the costs on the vertical axis.

State three other pairs of variables which might be used in a business graph.

Difficult decisions

Sometimes, it is a little difficult to decide which is the independent and the dependent variable.

For example, if you were considering the amount that a firm spent on advertising and its sales revenue, it might be difficult to say which is the independent and the dependent variable because so many other variables are involved, such as

- whether the market is growing or getting smaller
- whether consumers have a lot, or very little, money to spend
- how much firms in the same market spend on advertising etc.

In practice, advertising would usually be taken as the independent variable, as advertising budgets are fixed, and sales revenue is the dependent variable. Similarly, if you were considering the number of goods produced by a factory and the costs of doing so, which would be the independent variable? It is difficult to say: it depends on the point of view.

Single Line Graphs

In practice, the output, or the number of goods produced, is usually taken as the independent variable and is, therefore, shown on the horizontal axis. The costs, or dependent variable, are shown on the vertical axis.

On the whole, whenever we want to measure the effects of change on a variable, we put it on the vertical axis. So, in the previous example, we want to know what effects output has on costs, so costs go on the vertical axis.

In your view, which would be the independent and the dependent variable in the following pairs of variables:

> distance travelled by sales staff/amount spent on petrol
> house prices/wages
> number of cigarettes sold/deaths from lung cancer

Distortion

Line graphs can sometimes give a distorted, or misleading, impression of the facts. This can occur through lack of skill or, occasionally, through deliberate deception. Look at Figures 7.5 and 7.6 below. For what business purpose might the graph in Figure 7.6 be used? Explain the reasons.

In a line graph, the vertical scale should normally start at zero, as in Figure 7.5 below. The curve will then give a true picture of the rise in sales.

If the vertical scale starts at some higher figure, the curve will be greatly distorted, as in Figure 7.6 below, which shows a far more dramatic rise in sales, because the vertical scale starts at 18 million.

Figure 7.5
The sales of Brand A presented in a true way

Figure 7.6
Sales presented with the deliberate intention to deceive

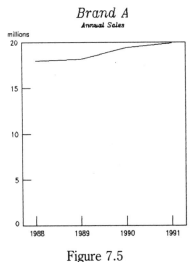

Figure 7.5

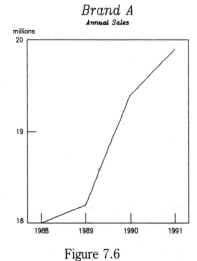

Figure 7.6

32 Graphs and Charts

The scale has been compressed to the greatest possible extent so that the increase in sales seems enormous. The more the vertical scale is compressed, the greater will be the distortion of the facts.

A graph like Figure 7.6 might be used deliberately in publicity or advertising to deceive consumers or investors.

Exceptions

There are times when it is reasonable to start the vertical scale with some other figure than zero. The price of company shares is a good example.

Let's say an imaginary company, Trueshaw plc, was formed 23 years ago. The shares were first issued at a nominal price of £1 each, as offered in the company prospectus. However, the company has been so successful that the shares have been bought and sold on the Stock Exchange for much higher prices for many years. Investors are not interested in the nominal value of the shares, which is the £1 they cost when they were first issued 23 years ago. What they want to know is the price of the shares month by month over the whole year.

Figure 7.7 below provides investors with the facts they want. Line graphs, like Figure 7.7, are frequently used in the financial pages of newspapers and in business magazines. They provide a realistic picture of the facts for the intended readership.

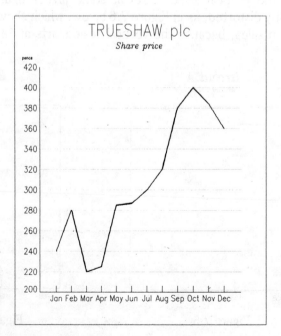

Figure 7.7
Share prices are an exception to the rule that the vertical axis should start at zero

However, if you included a graph of this kind in a coursework assignment, it would be advisable to state clearly in the adjoining text the reason for not starting the vertical scale at zero.

If your software permits, you could also break the vertical scale with a jagged line or a double slash, like this //, to indicate that part of the scale has been left out.

Hand-drawn graphs

Single line graphs are fairly easy to construct by hand. You will need squared graph paper, a ruler and a pen or pencil.

Decide on the scales you are going to use for the vertical and horizontal axes, or lines, before you start drawing the graph. (This is done automatically by software.) You want to use the space to the best advantage, so that the curve is not compressed into one small section of the graph as in Figure 7.5 above. You can break the vertical axis with a jagged line or a double slash (like this: //) to show that part of the scale has been left out. So, in Figure 7.5, you could draw the vertical axis up to 5 and then use a jagged line and make the next division 15, which would mean that you do not have such a top heavy graph (see Figure 7.8).

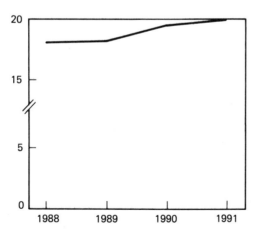

Figure 7.8

When you have drawn both axes, plot the coordinates carefully on the graph. You can mark them with a circle like this o. Join the coordinates with a straight line. Don't forget to label the axes, where necessary, and to add a title, which should be bigger than any other writing on the graph. Include a note of the source of your information at the bottom of the graph.

Activities

1

Use Figure 7.7 to answer the following questions:

(a) In which month of the year did Trueshaw shares reach their lowest level and what was their value in £s and pence?
(b) When was the sharpest increase in price?
(c) If an investor bought 2000 shares in March, what profit (excluding broker's charges) would he/she have made if the shares had been sold in

    (i) October
    (ii) December?

2

While carrying out a local enquiry, you have discovered the following information. Create a single line graph to illustrate the changes in the number of new businesses.

Number of new businesses started in the area

Year	Number
1987	84
1988	109
1989	99
1990	73

Unit 8 Step Graphs

Step graphs provide another way of showing a trend over a period of time. Instead of using a curve as in a line graph, the trend is shown as a series of steps, as in Figure 8.1 below.

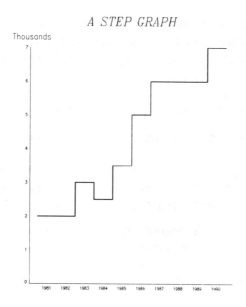

Figure 8.1
Step graphs are like a flight of (very uneven) steps

Note that the steps are of different sizes. Some are broader than others, showing that there has been no upwards progress for some time. How many broader steps are there in Figure 8.1 above?

You may have noticed that the years on the horizontal axis do not coincide with the beginning of each step. That does not matter in a step graph. It is the *width* of each step which is important, not the values on the vertical scale.

Upwards steps

Step graphs should only be used when there is some general upwards (or downwards) trend which can be shown in a series of steps, even though there may be a few tumbles on the way! They are ideal when you want to show how some barrier has been broken, or some high point has been reached, such as the steps by which a company reached an annual

36 *Graphs and Charts*

turnover of £1000 million. For example, they could be used to show the steps towards.

- the present number of homes in Britain or in your own area
- the present level of consumer credit
- making your first million!

Downwards steps

Step graphs are equally useful for showing a downwards trend, that is, the steps by which a present *low* point was reached. For example, the step graph in Figure 8.3 below shows how the building of tankers has declined in the United Kingdom

Study Figures 8.2 and 8.3 below and then answer the following questions.

(a) What are the main differences between a line graph and a step graph?
(b) Which graph has the bigger impact? Explain your reasons in full.
(c) What, in your view, has caused the decline in the building of tankers in the United Kingdom?

If you have the right kind of data, you could use an occasional step graph to add variety. They are very easy to create with most software. The line graph in Figure 8.2 below was changed into the step graph in Figure 8.3 with a single keystroke!

Figure 8.2
A line graph shows the downwards trend in the building of tankers (source: Annual Abstract of Statistics, *1986)*

Figure 8.3
A step graph shows the steps by which a certain point was reached (source: Annual Abstract of Statistics, *1986)*

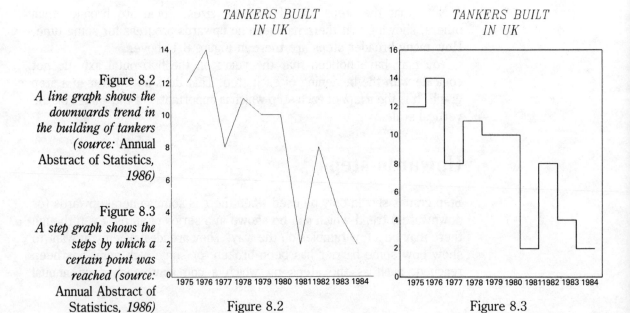

Figure 8.2

Figure 8.3

Hand-drawn graphs

Step graphs are not particularly easy to create by hand. If you decide to use one, you would have to mark out the points on the graph, just as you would with a line graph. Then, you would have to use these points as the central point for each of the steps. All of the steps would have to be of the same size, except when two (or more) successive values are the same.

Activities

1

Look in the latest volume of the *Annual Abstract of Statistics* in your main public library, and find some data which would be suitable for a step graph. Create both a line graph and a step graph using the same data.

2

State two examples of when a step graph might be used.

Unit 9 Multiple Line Graphs (1)

Multiple line graphs have two (or more) curves instead of only one like single line graphs. They are used to show different sets of data in the same graph.

Again, you have to be very careful that the data will produce a well-designed graph. Look at Figure 9.1 below. State what is wrong with it. What other kind of graph or chart would it have been better to use? Explain the reasons.

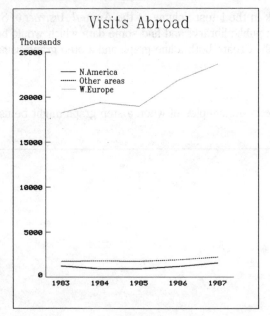

Figure 9.1
A multiple line graph shows several sets of data

Curve styles

To distinguish one curve from another, different styles must be used for each. In Figure 9.1 above, a dotted line is used for the top curve, dashes for the middle curve, and a solid line for the bottom curve.

Software usually provides other styles as well, such as a combination of dots and dashes. You should choose styles for each curve which are as different from each other as possible.

Legends, or keys, can be used to show the meaning of each curve. You should make the legends as short and as clear as possible.

Most software 'writes' the legends at a fixed point in the graph, for example, the top left-hand corner. Sometimes, the legends may lie right across one of the curves. It is usually very easy to move the legends to a different part of the graph with a couple of keystrokes.

Instead of using legends, you could label the curves with the actual words instead, that is, in Figure 9.1 above, 'North America', 'Other areas', 'Western Europe'. This looks neater and makes it easier for the reader to understand the graph; but it takes more time than using legends.

Dependent variables

Like single line graphs, multiple line graphs must always show a direct relationship between the variables. There is normally no problem with the independent variable, which is shown on the horizontal axis, as that is usually time. However, you must be extremely careful with the dependent variables on the vertical axis.

The dependent variables must always be of the *same kind*. It must be possible to compare them with each other. It would be all right to create a graph showing changes in a firm's sales and cost of sales over a period of time. The two dependent variables – sales and cost of sales – are closely related. However, it would be impossible to use the value of a country's imports and the costs of education in the same graph, as they are not closely related in any way.

The dependent variables must also be measured in the *same units*. It is all right to use sales and costs of sales together in the same graph, as they are measured in the same units – £s. However, the numbers of unemployed people and the cost of social security payments could not be used together, as one is measured in numbers and the other in £s.

State two other pairs of dependent variables which could be shown together in the same graph.

Index numbers

Numbers, percentages or £s can be used on the vertical axis. Sometimes, to avoid cluttering up a graph with a mass of large numbers, which people find difficult to understand, index numbers are used instead, as in Figure 9.2 below.

If you used the actual figures for retail sales, they would run into billions of £s. Instead, the volume of sales for each branch of retailing in one particular year (in this case, 1980) is given the standard number of 100. The volume of sales in later years is then worked out in relation to that standard number.

For example, the volume of sales for the top curve, household goods, was 100 in 1980. By 1981, the index number had increased to 102, and by

Graphs and Charts

1987 it had risen to 162. In other words, there was a 62% increase in the volume of sales between 1980 and 1987.

On the other hand, the volume of sales for the bottom curve, other non-food, actually fell by 1% to an index number of 99 in 1981, and rose to only 113 by 1987, a total increase of only 13% in the seven years.

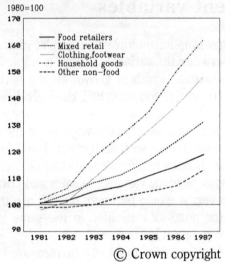

Figure 9.2
A baseline may be included in a line graph when indices are used on a vertical axis (source: British Business, *7 October 1988)*

Note that a solid baseline is used for the standard figure of 100 on the vertical scale so that falls are clearly shown as well as rises.

The use of legends make it rather difficult for the reader to see at a glance which curve is which. When there are more than three curves, it is probably better to label them instead.

Look at Figure 9.2 again and answer the following questions:

(a) Which retailers' sales fell in 1981?
(b) Which kind of goods have shown the greatest rise during the period?
(c) What has happened to the volume of food sales during the six years?
(d) What was the approximate rise in the volumes of sales of clothing and footwear in the whole period?

Hand-drawn graphs

It is possible to create multiple line graphs by hand; but you will have to take great care to make the different styles for each curve consistent throughout their length. Do not use legends. Label the curves neatly at the end, instead.

Multiple Line Graphs (1)

Activities

1

The index figures for the *value* of retail sales from 1985 to 1987 were as follows:

1980 = 100	*1985*	*1986*	*1987*
Food retailers	147	157	167.7
Mixed retail	146	159.3	172.9
Clothing and footwear	155	170	186
Household goods	157	175	192
Other non-food	154	165	180

Source: *British Business* 7 October 1988

The actual sales in £ million in 1980 were:

Food retailers	22,859
Mixed retail	10,945
Clothing and footwear	5,413
Household goods	9,234
Other non-food	9,926

(a) Construct a multiple line graph for the value of retail sales from 1985 to 1987 using the index figures above.
(b) What were the actual sales for the five kinds of retailer to the nearest billion pounds in 1987?
(c) Using the information, above and that in Figure 9.2, state which branch of retailing you would enter if you were setting up your own business. Explain your reasons in full.

2

Look carefully at the information below and then answer the questions which follow.

Learite plc makes machines at the two factories it owns in Liverpool. Its Managing Director is John Wilson who has written a report for employees. Part of this report is given overleaf.

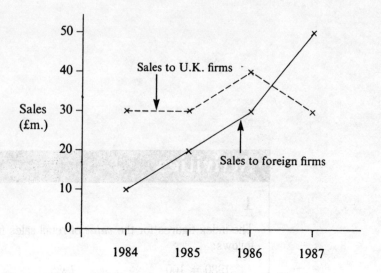

Figure 9.3

"You can see from the graph that Learite's sales have continued to increase. I am proud of the way in which Learite's sales have helped the U.K. balance of payments. We have faced many problems when selling to other countries and we have received much help from other firms and the Government.

1988 should be an exciting year and we hope to set up a factory in France. This will be the first of many factories in other countries."

(a) What was the value of Learite's sales to foreign firms in 1985?
(b) What was the value of Learite's TOTAL sales in 1986? Show your working.
(c) Is John Wilson's statement correct that "Learite's sales help the U.K. balance of payments"? Give reasons for your answer.
(d) Give brief details of the problems that Learite might face when selling to other countries.
(e) Why does the Government help those firms who wish to sell to other countries?
(f) Why do you think Learite wants to set up factories in other countries?

(Source: Southern Examining Group, Commerce Paper 2, Summer Examination 1988)

Unit 10 Multiple Line Graphs (2)

The multiple line graph really comes into its own when the curves cross each other and neatly fill the whole of the graph, as in Figure 10.1 below.

No other kind of graph or chart could show the steep fall in the building of Council homes and the upwards trend in the building of private homes in such a dramatic way. Multiple line graphs are ideal for showing contrasts of this kind, when one curve is falling and the other is rising.

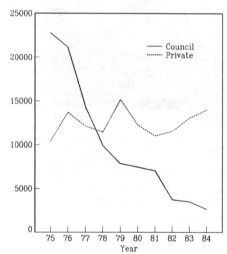

Figure 10.1
Curves that cross each other are the sign of a good multiple line graph (source: Annual Abstract of Statistics, *1986*)

Layer charts

When there is no great contrast of this kind, and the curves are stacked one above the other with wide open spaces in between, it is better to use a layer chart – if your software permits. Figure 10.2 below shows an example of a layer chart.

Layer charts (which are sometimes called band curves or strata charts) have one great advantage. The contrasting shading in each layer makes them much more interesting to look at than an ordinary multiple line graph.

Layer charts give a quick impression of the main facts, but they are rather difficult to read, especially with the middle layers. It makes it easier for the reader, if you include the vertical axis on both sides of the chart.

44 *Graphs and Charts*

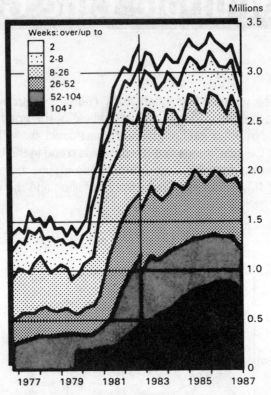

Figure 10.2
Layer charts use contrasting shading (source of chart: Social Trends 18*)*

1 Figures up to October 1982 relate to the registered unemployed. From October 1982 the figures are on the new basis (claimants). From April 1983 some men aged 60 or over did not have to sign on at an Unemployment Benefit Office to receive the higher rate of supplementary benefit and national insurance credits. Between March and August 1983 the number affected was 162 thousand, of whom about 125 thousand were in the over 52 weeks category.

2 Data are only available from October 1979.

© Crown copyright

What other kind of chart could have been used instead of a layer chart for Figure 10.2?

Combination charts

With most software, you can create some extremely professional-looking charts by combining, or putting together, different kinds of charts and graphs. Figure 10.3 below shows the result of combining a multiple bar chart and a single line graph. The curve shows the actual percentage profits made by a firm during the four quarters of a year. The bars show the profit forecasts of the firm's Northern and Southern divisions.

(a) Which division, the Northern or the Southern, is better at forecasting profits?

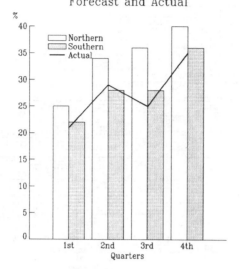

Figure 10.3
A bar chart and a line graph can be used in the same chart

(b) Which quarterly forecast was
   (i) most accurate,
   (ii) most inaccurate
(c) Give two other examples of when a combination chart of this kind could be used.

Note that the bars have different shading to make it clear which is the Northern and the Southern; and that a thick curve has been used to make it stand out more as it crosses the bars. Why wasn't a solid fill used for one of the bars?

Hand-drawn graphs

Although you can draw a multiple line graph by hand, it is not advisable to attempt the more complicated layer and combination charts.

Activities

1

A brick manufacturer wishes to assess road and rail transport as means of delivering loads of bricks to customers nation-wide. Use the graph of Figure 10.4 and your own knowledge to answer the following matters raised by the manufacturer.

(a) (i) Which is the cheaper method of transport for a 100 mile journey, A or B?

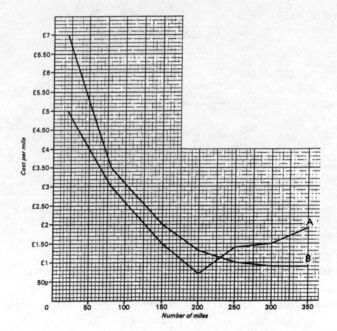

Figure 10.4
(Source: London and East Anglian Group, Business Studies, Specimen Paper 1)

    (ii) Which is the cheaper method of transport for journeys over 250 miles, A or B?
    (iii) At what distance does it cost the same by A and B?
(b) What is the difference in *total* cost between A and B for a journey of 200 miles?
(c) Identify the form of transport called A and state your reasons.
(d) Identify the form of transport called B and state your reasons.

2

Look back at Figure 10.1 and then answer the following questions:

(a) In which year did the building of private homes in Scotland reach a peak?
(b) What has happened to the number of Council homes built during the whole period? What, in your view, has been responsible for the change?
(c) Explain in full what effects all the changes are likely to have had during the whole period on:

    (i) waiting lists for Council homes
    (ii) building firms.

3

Explain all the advantages and disadvantages of a layer chart compared with a component bar chart.

Unit 11 Histograms

A histogram is very different from any other chart or graph we have considered so far. Instead of providing information about numbers or percentages, it shows the *frequency*, or number of times, that different values of a variable occur.

Before a histogram can be created, the data must be collected. Let's take a very simple example of how this might be done.

Say you wanted to find out how much money 16-year-olds in your area had to spend each week. You would have to find out how much they earned by part-time jobs, how much interest they received from any savings, how much money they got (per year) in presents – and what pocket money they received.

You could ask a few of your friends about their pocket money and obtain the following answers:

$$£1.50 \quad £5 \quad £2.50 \quad 50p \quad £1 \quad £1.30$$

That data would not get you very far. There is not enough to provide a true, general picture for the area.

You could find the average amount (or arithmetic mean) by adding all the amounts together and dividing by the total number, that is:

$$\frac{£11.8}{6} = £1.97$$

But that wouldn't be much better. The average, £1.97, is an amount that no one actually receives.

What you want to find out are the general levels of pocket money among 16-year-olds in your area. So you would have to ask many more of your friends, and other people you know, to obtain more data.

Tallies

Before you did this, it would be a good idea to make out a table, called a tally or an array, like Figure 11.1 overleaf. This will make it easier for you to keep a record of the data.

Take the lowest amount anyone can receive – which is zero – and the highest amount they are likely to receive. Then divide these into equal groups, or classes. Note that the classes are given in pence: 0–99p, 100–199p.

48 Graphs and Charts

POCKET MONEY

AMOUNT		FREQUENCY
0-99p	IIII	4
100-199p	IIII IIII III	13
200-299p	IIII IIII IIII II	17
300-399p	IIII IIII I	11
400-499p	III	3
500-599p	I	1

Figure 11.1
A tally provides a simple way of grouping data into classes

Note that a mark is put against each amount as it is recorded. When you have four marks in the same class, you put a diagonal stroke through them, instead of a fifth mark. This makes it easier to count up at the end, as each completed block contains 5 marks.

Frequency distribution

From your tally you could then produce a table, like the one below.

POCKET MONEY

Amount received per week in pence	Number of people
0–99	4
100–199	13
200–299	17
300–399	11
400–499	3
500–599	1
	49

This is called a grouped frequency distribution, because the different amounts of pocket money are divided into groups, or classes (0–99p etc), and because it shows how often this class occurs, or its frequency. The class with the greatest frequency is called the modal class. Which is the modal class in the table above?

Histograms

This information could be used to create a histogram, as in Figure 11.2 below

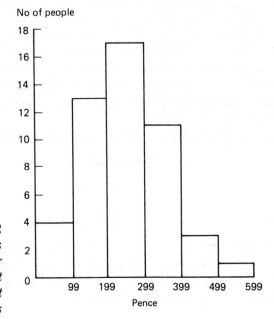

Figure 11.2
The histogram shows the frequency, or number of times, that a range of pocket money occurs

It is very easy to recognise a histogram

- the bars are always joined together, unlike those in a bar chart
- the scale on the horizontal axis, or line, must always form a continuous series, with no values omitted, as in Figure 11.2 above.
- the vertical axis always shows frequency.

Figure 11.2 gives only an approximate, or rough, idea of the frequency. What are the reasons?

Stem-and-leaf diagrams

If you wanted to present a much more accurate picture, you could use a stem-and-leaf diagram to illustrate the results of your survey. Instead of grouping the data into classes, you just collect the data, that is, the amount of pocket money that each person receives. Say the amounts were:

£2.50	£1	30p	£1	£2
£2	£3	£4	70p	£2.50
£3	£3.50	£3.50	£3	£3
50p	£5.50	£1.20	£1.70	£1.70
£2	£2	£2	£3	£3.50
£4	£1	£1.70	£1.70	£2.50
£2.50	£2.50	£3	£3.50	£4.50
70p	£1.20	£2	£2	£2.50
£2.50	£3.50	£2	£1.70	£1.70
£1.20	£2.50	£2.50	£1.20	

In its present form, this is just a confused jumble of numbers; but it can quickly be put into shape with a stem-and-leaf diagram

In this case, we will take 100s (or £s) as the stem, and tens (or 10p) as the leaves. So, if we take the first amount in the table above, £2.50, the stem would be 2 and the leaf would be 5. The next amount (reading across the columns) is £1. Therefore, the stem would be 1 and the leaf would be 0. The next amount is 30p. The stem would be 0 (as there are no pounds) and the leaf would be 3.

In a stem-and-leaf diagram, these would be entered as:

Stem-and-leaf

2	5
1	0
0	3

To make a complete stem-and-leaf diagram, all the amounts are dealt with in the same way, entering the £s in the stem column and 10ps (or 0) in the leaf column. The first six lines of the table above would appear in the following way in a stem-and-leaf diagram:

Stem-and-leaf

5	5
4	00
3	00550005
2	50050005
1	00277077
0	375

Copy out the diagram above and complete it by filling in the numbers from the rest of the table above, starting with £2.50 at the beginning of the seventh line.

As you can see a stem-and-leaf diagram provides more precise, or accurate, information than a tally and shows more clearly how the numbers cluster in each batch or group. The information could be used to create a vertical line chart as in Figure 11.3 below. (There are not 49 lines, as many people receive the same amount of pocket money.)

What is now the modal value? Why is it different from the modal class in the histogram?

If you joined the top of each stick, you would produce a much more accurate frequency curve. Vertical line charts (or stick charts as they are sometimes called) have other uses, as we shall see in Unit 12.

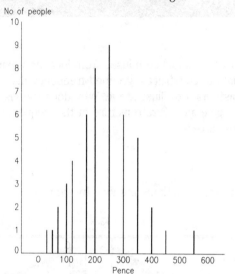

Figure 11.3
Vertical line charts provide very accurate information

Main uses

Any of the charts discussed in this unit can be used when frequency is involved. They can be used for information relating to the frequency of such items as:

- classes of wages
- number of people in households
- ages at which people marry
- household income divided into classes.

Hand-drawn charts

A vertical line chart is probably the easiest to draw by hand (see Unit 12).

Histograms are more difficult. You would have to use the same procedure as for step graphs, described in Unit 8, drawing complete rectangles instead of open-sided steps.

Activities

1

Collect prices of detached houses from local newspapers in your area. Use the data to construct a grouped frequency distribution, and then create a histogram to illustrate it. Why does the histogram not necessarily give an accurate picture of the frequency of actual house prices in the area?

2

Histograms are sometimes put side by side to form a population pyramid.

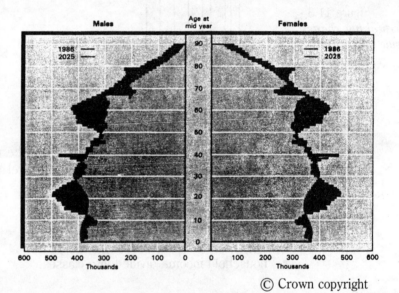

Figure 11.4
(*Source of chart:* Social Trends 18)

© Crown copyright

(a) Why is it called a pyramid?
(b) Describe the main changes which are likely to occur in the population structure between 1986 and 2025.
(c) What effects, in your view, are these changes likely to have on the business world?
(d) How effective is a population pyramid in conveying information?

Unit 12 Vertical Line Charts

Vertical line charts can be used to present other information as well as frequency. They are especially useful if you want to show changes in a variable in great detail. For example, they could be used to show exact changes over a period of time in:

- exchange rates
- a firm's sales
- the number of shares bought and sold on the Stock Exchange
- the rate of inflation.

Inflation

The rate of inflation shows the general rise in prices over a period of time. It is based on the prices of about 350 goods and services, which are collected monthly. Usually, the rate of inflation is expressed as a percentage. An annual, or yearly, rate of inflation of 5% would mean that something which cost £1 to buy a year ago would now cost £1.05. What would it now cost if the rate of inflation over the year had been 10%? What would something which cost £45 a year ago now cost if the annual rate of inflation had been 6%?

The rate of inflation can vary greatly from year to year, and even from month to month. For that reason, you have to be very careful in choosing a chart or graph to illustrate it.

Why would it be wrong to use a line graph like the one in Figure 12.1 overleaf?

Figure 12.1 doesn't provide a true picture of the facts, because it gives the rate of inflation for only four years out of the thirty. It does not show what happened in the years between. If you tried to read off these values, your answers would be hopelessly wrong. (Remember what was said in Unit 7 about the dangers of reading off intermediate values in a line graph?)

Of course, you could create a line graph which showed the rate of inflation for each year, but the vertical chart in Figure 12.2 overleaf is better for two reasons. It shows the changes in the rate of inflation much more dramatically. Furthermore, there is no temptation to read off monthly values, as there might be with a line graph, since the sticks are separate.

Most software makes it very easy to create a vertical line chart. You can have dozens of lines in a chart. The more there are the better. Not only will the chart be more accurate, but the curve, formed by the top of the lines, will stand out more clearly.

54 Graphs and Charts

*Figure 12.1
A line graph with 10-yearly intervals gives a very false view of the rate of inflation (source: Department of Employment, Retail Prices Indices)*

*Figure 12.2
A vertical line chart gives detailed information about inflation (source: Department of Employment, Retail Prices Indices)*

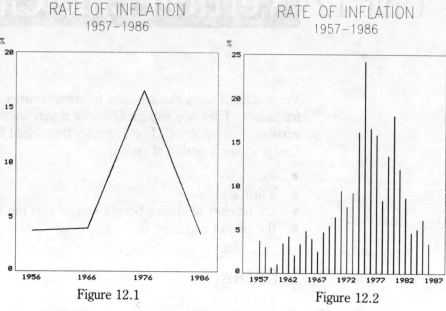

© Crown copyright

Hand-drawn charts

If you draw a vertical line chart, you must be careful to make the lines parallel, and to leave an equal space between each one.

Activities

1

Look at Figure 12.2 again and then answer the following questions:
(a) What was the highest level of inflation and in which year did it occur?
(b) In which year did inflation reach its lowest point?
(c) Describe in detail what has happened to inflation since Mrs Thatcher came to power in 1979. Explain from your own knowledge the main reasons for the changes.
(d) In which period was the rate of inflation generally low?

2

Find a table in the *Annual Abstract of Statistics* which shows great variations in the values of a variable over a long period of time. Use the figures to create a vertical line chart.

Unit 13 Scatter Graphs

Line graphs are used to *show* the relationship between variables. Scatter graphs are used to help *decide* whether there is an association, or correlation, between two variables, that is, whether changes in one variable are related to changes in the other. They are not used to present facts, but help to answer questions.

For example, we might want to find out whether there is any relationship between the amount of money that firms spend on capital equipment, such as expensive tools and machinery, and their sales revenue.

We could collect data from a number of firms and create a scatter graph (or diagram) like Figure 13.1 below.

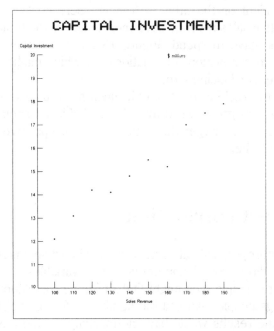

Figure 13.1
The dots on a scatter graph help to show whether there may be a relationship between two variables

Perfect correlation

The pattern of dots, or points, on the graph show the degree of relationship between the two variables.

If all the dots form a straight line, there would be a perfect correlation, or a very definite association between the variables. If, on the other hand, they were scattered about all over the graph, there would be no correlation.

Look at Figure 13.1 above. It is obvious, at a glance, that if the points were joined together, they would not form a straight line. Now place a ruler on the graph so that it joins the first and the last point. If you drew a line between these two points, five of the other points would be very close to the line, and none of the other three would be very far away.

Positive correlation

Figure 13.1 is said to show a positive correlation. In other words, there seems to be a very strong relationship between capital investment and sales revenue, that is, the rise in values seems to be related. (There can also be a negative correlation, when a fall in values appears to be related.)

A scatter graph does not show cause and effect, or that one variable is dependent on the other. After all, there could be many other variables affecting sales revenue even more, such as the total amount of money that people have to spend. Figure 13.1 above simply shows that there is a positive correlation, or relationship, which might be worth investigating further, or looking into.

In the real business world, much more data would be obtained before the scatter graph was drawn. Instead of just 10 points, there might be 100 or more. The larger the sample, the more positive the relationship would seem to be.

Direct correlation

A scatter graph should only be used when it seems reasonable that there is some direct correlation between two variables.

For example, you might find a positive correlation between the number of dogs people own and the number of foreign holidays they have. If you did, the results would have no meaning, as there is no direct connection between the two variables.

On the other hand, the number of books people read and the age at which their education ended would be a suitable subject for a scatter graph, as it seems likely that there is some direct correlation.

State two other pairs of variables which would be suitable for a scatter graph.

Same axes

With many software packages, you can select a scatter graph with the tap of a single key. The data is entered in the normal way.

Note that the axes in the scatter graph are just the same as those in an ordinary line graph. Only the relationship between the two variables is being considered, so the vertical axis doesn't always have to start at zero.

For a similar reason, the points on the graph are not joined by a curve, because one variable is not necessarily dependent on the other.

For example, you might find a positive (or negative) correlation between the sales of ice cream and swimming suits; but this would only show that they are related. In fact, both of them are dependent on the daily temperature, which is the independent variable. If it is hot and sunny, the sales of both ice cream and swimming suits will rise.

Use in enquiries

Scatter graphs can be very useful when you are making your own enquiries. Say you had collected data about two variables in the hope that you would find some association between them. In the end, however, you discover that you have been going up a blind alley. There is no correlation between them at all.

Instead of wasting all your work and research, you could use a scatter graph to illustrate the lack of any correlation. You would, of course, have to explain in the text why you had chosen the variables and why you thought they might be related.

Hand-drawn graphs

It is just as easy to draw scatter graphs as line graphs. Draw the axes in the same way and simply plot the points from the coordinates, but do not join them with a curve. Make sure that you use a dot that is big enough to see; or, if you prefer, you could use a small circle.

Activities

1

Create scatter graphs of different kinds, making up your own data.

2

During a local enquiry, you have collected the following information about six firms in the same industry:

Advertising Expenditure/Turnover

£s

Turnover	10,000	18,000	23,000	36,000	45,000	55,000
Advertising expenditure	1,500	2,200	2,800	3,200	4,200	5,000

(a) Create a scatter graph based on this data.
(b) What kind of correlation does it show?
(c) What other variables would you have to take into account before you could decide whether or not there was a positive correlation?

Unit 14 Straight Line Graphs

Straight line graphs are used when the value of each variable changes at exactly the same rate, such as an hourly wage (without overtime) and the number of hours worked. Say the hourly wage is £3. The total wage for 10 hours' work would be £30 (£3 × 10 = £30). For 20 hours, it would be £60 – twice as much. What would the total wage be for 15 hours?

Value added tax (VAT) and prices is another example. VAT is added on at the standard rate of 15% for most goods and services. The price of a personal stereo, without VAT, might be £20. The VAT on £20 is £3, that is:

$$\frac{£20}{100} \times 15 = £3$$

Therefore the price of the personal stereo in the shop would be £23 (£20 + £3 VAT). Something which cost £40 without VAT would cost £46 in the shop. What would be the shop price for something which cost £80 without VAT? (In practice, VAT is collected at each stage of production, from the manufacturer, the wholesaler, and the retailer; but it is the consumer who actually pays the total bill over the shop counter.)

As the rate of VAT is always the same, the relationship between the price without VAT and the price including VAT can be shown in a straight line graph as in Figure 14.1 below.

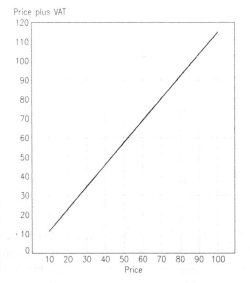

Figure 14.1
Straight line graphs are used when the values of variables change at exactly the same rate

Direct variation

The scales in Figure 14.1 above can be used either for pence or £s, though the same units must be used on both axes. As values change at the same rate on both axes, intermediate values can be read off at any point.

This kind of relationship, when the values of the variables change at exactly the same rate, is known as a direct variation.

Similar graphs could be used to show the direct variation between:

- the current exchange rate of, say, £s and $s
- the number of customers at a one-price concert and turnover
- the output of the same kind of goods and sales revenue.

Let's say that a young person is making one kind of toy which sells for £10 (excluding VAT). Obviously, there is a direct variation between the number of toys made and the sales revenue:

50 toys would bring in £500
100 toys would bring in £1,000

How much would 200 toys bring in?

The relationship between output and revenue can be shown in a straight line graph as in Figure 14.2 below.

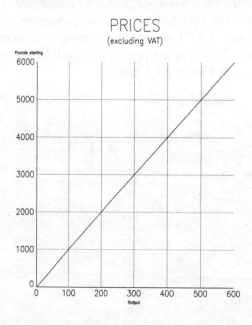

Figure 14.2
Output and sales revenue change at the same rate

This graph can easily be extended into a break-even chart, which is used to show when a business is likely to become profitable. Let's say that the fixed costs of the business, such as rent, business rates, electricity etc., were £2000 a year. These have to be paid whether a single toy is made or not, so a baseline is added at £2000 on the vertical axis, to represent the fixed costs, as in Figure 14.3 below.

On top of that, there are variable costs of £6 for each toy. These increase as more toys are produced. So, the variable costs for 100 toys would be £600, and for 200 toys, £1200. This line (shown as a dotted line in Figure 14.3 below) is drawn from the £2000 baseline, so that it shows the total costs, that is, variable costs *plus* fixed costs. The point at which this line and the revenue line meet is the break-even point.

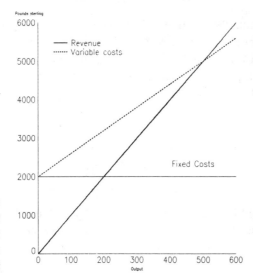

Figure 14.3
A break-even chart shows the point at which a business is likely to make a profit

Look back at Figure 14.3:

What would be the break-even point?
What would the turnover be at that point?
What would the financial situation be when 600 toys were produced?

Instead of using a chart to find the break-even point, it can be found by a couple of simple calculations. Deduct the variable costs from the selling price for each toy (or unit produced). In this case, it would be:

£10 (selling price) − £6 (variable costs) = £4

Divide the fixed costs (£2000 in this case) by the result of the first calculation (£4 in this case) to find the break-even point, that is:

$$\frac{2000}{4} = 500$$

Break-even charts can only give an indication of what is likely to happen. They do not provide a definite answer. Fixed costs may change: the rent may be increased; there may be a sudden jump in business rates; the interest rates on borrowed money may rise or fall. Variable costs are also liable to change. For example, if the toy-maker couldn't produce 600 toys in a year, he/she would have to employ someone, so the variable costs would alter.

Economic theories

Software can also be used to produce some graphs to illustrate economic theories. For example, a vertical line chart and a single line graph are used in Figure 14.4 below to show a demand curve – the quantity which will be bought at a particular price. The data for each curve must be exactly the same, that is, in this case:

8
7
6
5
4

The straight line graph will then fit neatly on the top of the stick chart – like a roof. Note that a thicker line is used for the line graph to make it stand out more.

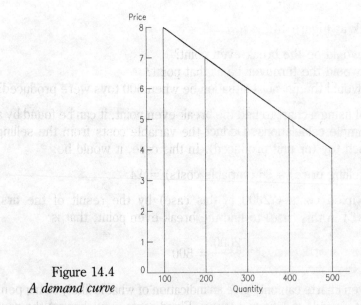

Figure 14.4
A demand curve

Supply and demand

A supply and demand diagram can be created in a similar way, as in Figure 14.5 below. In this case, the data on the two curves are in inverse order, with the values increasing on the first curve and decreasing on the second.

The point at which the curves cross is the equilibrium price, at which supply exactly equals demand. At a higher price, the number demanded will be smaller than the quantity supplied. By reading off the values in Figure 14.5 below, we can see that at a price of £500, 200 will be demanded and 550 supplied. At a lower price, demand will be greater than supply.

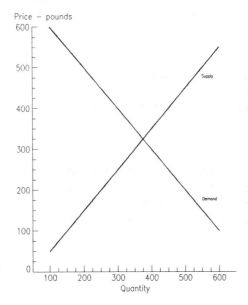

Figure 14.5
A supply and demand diagram

Use Figure 14.5 to answer the following questions.

(a) What is the equilibrium price?
(b) What is the quantity demanded at that price?
(c) How much is demanded and supplied at a price of £200?

Limitations

Although software can produce neater diagrams than you could by hand, it is probably not worth using it for simple diagrams like these. Furthermore, software usually produces only straight lines, and not the curved lines which are used in many diagrams illustrating economic theory.

Hand-drawn graphs

Most of these graphs and diagrams can be drawn fairly easily by hand. Do not attempt to use grid lines, unless you are certain that you can draw them neatly.

Activities

1

Find the current exchange rate of the £ and the dollar in the financial pages of a newspaper. Create a straight line graph based on this data.

2

A small business had fixed costs of £4000. Its variable costs for each unit it produced was £29 and the selling price was £40. Create a break-even chart. State at what point the business is likely to make a profit.

3

Choose a popular chocolate bar. Ask your friends how many more they would buy each week if the price were reduced by 2p, 4p and 6p, and how many fewer they would buy if the price were increased by the same amounts. Use this data to create a supply and demand diagram.

Unit 15 Pictograms

Figure 15.1
Pictograms are often used to present information in real life

Information is now very often presented in pictograms, like those in Figure 15.1 above. What do the symbols represent? Where would you expect to find them? Name three other pictograms you might see in everyday life.

Pictograms use symbols or little drawings to represent a thing, like a castle; a person, like a woman; or an idea, like not washing clothes, but dry-cleaning them instead.

They are called pictograms because they are *picto*rial dia*grams*.

Pictograms are the oldest form of communication. Thousands of years ago, primitive men and women painted them on the walls of the caves where they lived to show the kinds of animals they hunted. Ancient civilisations, such as the Maya, Aztec and Egyptian, used pictograms (or hieroglyphics), instead of words, in their writing. They are one of the simplest means of communication and the most easy to understand.

Business charts

Pictograms can also be used in business charts. Let's say you are investigating inland transport, and found the following information in the reference book *Social Trends*:

Passenger transport

Billion passenger kilometres travelled

	1961	1971	1981	1986
Air[1]	1	2	3	4

[1] Domestic scheduled journeys only, including Northern Ireland and the Channel Islands

Source: *Social Trends 18*.

66 *Graphs and Charts*

You could change this table into a pictogram quite easily by using two symbols, one for a plane and another for a passenger as in Figures 15.2 below.

Figure 15.2
A pictogram can be used to show the increase in air travel (source: Social Trends 18)

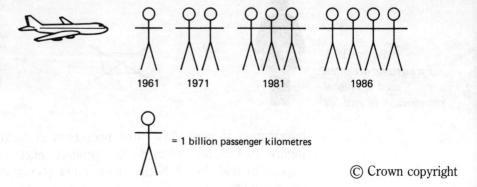

© Crown copyright

Graphics library

Pictograms provide one of the most dramatic ways of presenting information. However, there is one snag. You will have to draw them by hand unless you are using advanced software, such as a desk-top publishing packages. If you are, you may find there are enough symbols in the software's graphics library for your needs.

For example, the graphics library might contain a little drawing of a television screen. This could be used in a business chart for a number of different purposes. It might be used to represent changes over the years in:

- the number of television sets manufactured
- the number of TV sets in the whole country
- the total number of television viewers.

Make a sketch of a visual display unit, or computer screen, which will clearly distinguish it from a television screen. For what purpose could it be used in a business chart?

Symbols

To create a pictogram, you must first choose the right symbols. There are a number of ways in which this can be done. The easiest way is to use a simple sketch of the subject itself, as in Figure 15.3 below.

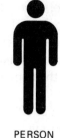

Figure 15.3
Simple sketches can tell the story

PERSON   LETTER   CUP

Draw a symbol for a car and state how it might be used in business charts. Another way is to use clothes. These can be used to represent:

- men and women (as on lavatory doors)
- different nationalities by using the clothes which are associated with that country.
- occupations by using the uniforms or clothes that are normally worn at work.

Draw pictograms to represent a policeman and a Dutch person.

The tools or equipment that people use is another way of creating symbols. For example, an engineer could be represented by a spanner. (You may have used this method already by adding a truncheon to your sketch of a policeman.)

What symbols would you use to represent a handicapped person, a teacher, a business person?

Number of symbols

In pictograms, you must be very careful about the number of symbols you use. There were no problems with the pictogram about air transport in Figure 15.2 above. The numbers were ideal for creating a pictogram. Unfortunately, that doesn't happen very often.

Let's say that you were writing a report on the car industry. August is always a good month for sales as people rush to buy cars with the new yearly registration letter. 1988 was a record year for new car registrations. To illustrate the point, you want to use a pictogram based on the following information.

Motor vehicle registration in Great Britain

Thousands

Taxation class	August 1987	August 1988
Private and light goods	433.8	512.0

Source: *British Business*, 7 October 1988

First of all, you have to choose a symbol. The trade includes both cars and light goods vehicles. There is no symbol which will represent both, so you have to cheat a little and use a symbol for a car instead.

Now look at the numbers. Obviously, you couldn't have 433 and 512 symbols in the chart. If you rounded down the numbers to 43 and 51, there would still be too many symbols.

You could round the figures to the nearest 100,000, so that you had 4 car symbols for 1987 and 5 for 1988. The number of symbols would be suitable for a pictogram, but the information would not be very accurate. Rounding the figures to the nearest 100,000 would give an error of about 2% for 1988, but an error of nearly 8% for 1987.

Pictograms may be dramatic, but they can also be very inaccurate.

Solutions

There are three solutions to the problem: but none of them is very satisfactory.

(1) You could round the numbers to the nearest ten thousand which would give 4.3 and 5.1, and use only part of the car symbol for the decimal fractions, as in Figure 15.4 opposite.
(2) You could use only one symbol for each year and increase its size in proportion to the rise in numbers. Note that it is the total area of the symbol which must be increased. You can work out the area by finding the percentage increase, which, in this case, would be:

$$512.0 - 433.8 = 78.2$$

$$\frac{78.2}{433.8} \times 100 = 18\%$$

The problem with this method is that the reader may find it difficult to judge the real size of the increase. On the other hand, if there is a really big rise, so that the area of the symbol has to be doubled or trebled, it may give a misleading impression.
(3) Alternatively, you could use the actual numbers in the pictograms. Methods 2 and 3 are combined in Figure 15.5 opposite.

Pictograms are most useful when clear, simple symbols can be used and the data can be easily divided into whole numbers.

Pictograms 69

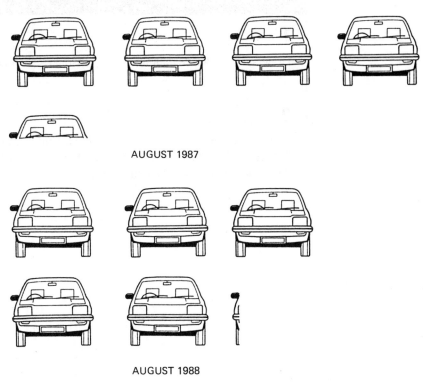

Figure 15.4
Part of a symbol has to be used for some data (source: British Business, *7 October 1988)*

= 100 000 cars

© Crown copyright

NEW CAR REGISTRATIONS

Figure 15.5
Using different symbols and actual numbers (source: British Business, *7 October 1988)*

433, 800

AUGUST 1987

512,000

AUGUST 1988

© Crown copyright

Activities

1

Draw symbols for restaurant meals, prisoners, cleaning services.

2

The number of Council homes built in the United Kingdom in 1974 was 121,017 and in 1984, 37,176 (*Annual Abstract of Statistics*, 1986). Use this information to create a pictogram.

Unit 16 Cartograms

Cartograms are simple maps which provide information about different areas. The information is usually presented in actual numbers and percentages as in Figure 16.1 below. They are used to make comparisons between different areas.

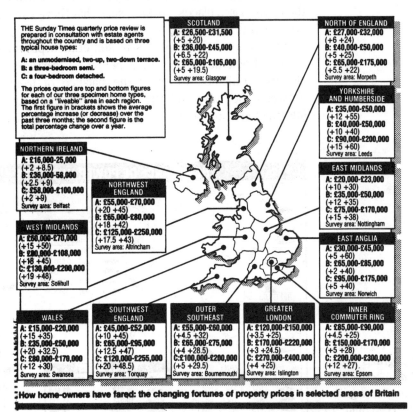

Figure 16.1 *Cartograms are often used to show regional house prices (source of cartogram:* Sunday Times, *9 October 1988)*

Copyright © *Sunday Times* 1988

Study Figure 16.1 above and then answer the following questions:

(a) How much would an unmodernised terrace house cost in Scotland and in Greater London?

(b) Which area, or country, has the lowest house prices? What in your view is the main reason?

(c) Where have house prices increased by the biggest percentage during the year?

(d) How much more would a semi-detached house cost in the South-west than in the East Midlands?

As you can see, a cartogram can provide a vast amount of information in an attractive way. The main danger is including too much information so that the reader becomes confused. How clear was the information in Figure 16.1 above? Was too much information included in your view?

By hand

Cartograms, like pictograms, have to be drawn by hand, unless you are using advanced software. Simply trace the map and add the information in the appropriate places, making sure that there can be no mistake about the area to which it refers.

Information in cartograms is sometimes provided by charts or pictograms. Look at Figure 16.2 below. A cartogram and pictograms are cleverly combined to show the balance of front-line military strength between the East and the West in Europe. Again, a large amount of information is presented in a very clear and dramatic way.

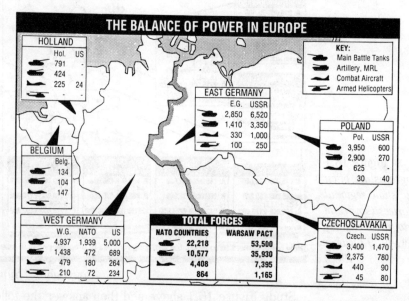

Figure 16.2
Pictograms can be included in cartograms (source of cartogram: Daily Telegraph, 19 October 1988)

Look at Figure 16.2 above and answer the following questions:

How many more tanks do the Warsaw Pact countries have than NATO?
Where are American forces mainly stationed on the Continent?
What are the proportions of Russian and East German combat aircraft in East Germany?

Most effective

Figure 16.3 is a particularly good example of a cartogram. It combines a simple, but effective, outline map with three-dimensional bar charts. It shows pictorially the effects that a motorway, like the M4, can have in persuading firms to move away from high rent regions to cheaper areas of the country.

At their best, cartograms are one of the most effective means of presenting information dramatically and economically. They can be used whenever geographical comparisons are being made.

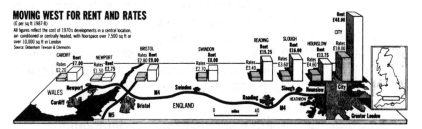

Figure 16.3
An imaginative use of bar graphs in a pictogram (source of cartogram: Sunday Times, *9 October 1988)*

Copyright © *Sunday Times* 1988

Main uses

Cartograms can be used in local enquiries to present such information as:

- the locations of main industries in an area, with actual numbers of people employed
- comparisons between business rates in different parts of the area
- unemployment in different parts of the region.

They can be used nationally to make comparisons between regions in such matters as:

- house prices
- wages and salaries.

They can be used internationally to make comparisons between countries in such matters as:

- national income per head
- balance of payments
- capital investment in industry.

74 *Graphs and Charts*

Activities

1

The graphics in Figure 16.4 provide information about the British water industry. Study them carefully and then answer the following questions:

(a) What are the three graphics called?
(b) What is water used for mainly in homes?
(c) What does RPI stand for in the graph?
(d) How does the rise in water bills compare with the rate of inflation? What was the reason in your view for the change in 1983–4?
(e) Which water authority makes the highest profit?
(f) Which has the highest debt?
(g) How satisfactorily do the graphics convey information to the reader? Describe any changes you would make and explain in detail your reasons for doing so.

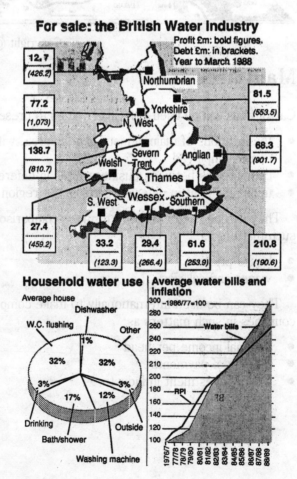

Figure 16.4
(*Source:* The Guardian, *24 November 1988*)

2

Create a cartogram showing garages in your area, including pictograms to show the main services they provide.

3

Use the following information to create a cartogram:

BUSINESS EXECUTIVE LIFESTYLES

%

	Yacht or other boat	Swimming pool
Great Britain	4.6	0.9
France	3.1	3.8
West Germany	3.9	7.5
Italy	8.6	1.6
Holland	9.0	0.2
Spain	8.0	28.6
Sweden	26.1	3.4
Switzerland	5.2	5.6

Source: *Daily Mail*, 21 September 1988 (Extracts)

Unit 17 Sources of Information

Before you can create charts and graphs, you must obtain information. It is very easy to obtain national information. The government publishes statistics about almost every topic you could ever think of, ranging from births to bank loans and from television to tourism.

The most useful books for obtaining statistics are:

Social Trends. This is published every year. It covers such topics as population, households and families, health and personal services, housing, transport and leisure. In addition to tables, it also contains text, which explains the tables, and graphs and charts.

The *Annual Abstract of Statistics* which deals with such matters as the National Health Service, education, employment, production, energy, building and construction, industries, transport, trade, government and local authority finance, banking and insurance. The information is provided in tables.

Both of these books are usually available in large public libraries. In addition, some public libraries may also have the *Monthly Digest of Statistics* which provides more up-to-date information about a selection of topics such as production, employment etc.

Latest information

The latest information on a variety of business topics may also be found in two official magazines:

> *Employment Gazette*
> *British Business*

Again, these magazines are usually available in bigger public libraries.

For the very latest information about business matters, you could look in the weekly magazine, the *Economist* and/or the daily newspaper, the *Financial Times*. Other daily newspapers which might provide useful information are the *Times*, the *Daily Telegraph*, the *Guardian* and the *Independent*. (These can also be found in most big public libraries.) The *Sunday Times* and the *Observer* are also worth studying. Trade associations may be able to supply you with statistics about their particular trade or industry. The reference librarian in your local public library will help you to find their address.

Local information

Some useful local statistics can be found in Council reports which are available in Council offices. These are also usually available in the reference department of public libraries.

The reference librarian may also be able to tell you about other collections of local statistics which may be useful.

Every 10 years, a census is made of all the people living in the country which records a mass of information about such topics as employment, education, age etc. Reports are published giving statistics for each region. These are usually available in bigger public libraries.

Unfortunately, this information is now rather out-of-date as the last census was held in 1981. When the results of the new census are available, after 1991, they will provide up-to-date local information.

Information about big local firms (public limited companies (plcs), whose shares are quoted on the Stock Exchange) can be easily obtained. Their report and accounts, which contain a mass of statistics about the firm's activities, are published every year. A copy can usually be obtained by writing to the firm.

Collecting data

In addition to using published statistics, you may have to collect some data yourself when you are making local enquiries.

Observation surveys are one of the easiest ways of collecting data. Let's suppose that you were writing a report about a pressure group which wanted the Council to build a relief road to reduce traffic noise and the risk of accidents. You could count the number of lorries, buses, cars and other vehicles which pass along the main street in the area. Using this data, you could then create a graph or a histogram to illustrate your report.

Of course, you would have to be careful to collect the right sort of data. If you carried out your count in the middle of the night, you wouldn't get a very true picture!

Equally, a count carried out during the rush hour might not be representative of the volume of traffic at other times of the day. The day of the week, the season of the year, and the weather could also influence your results.

Ideally, traffic counts should be carried out at different times of the day and in different seasons of the year to get a really true picture. However, you can still use the data you collect, even if you carry out only one count or census, so long as you mention its limitations in your report.

You might carry out similar observation surveys about:

- the numbers of people who look at a poster in the street

- how many people buy loss leaders in a supermarket (having first asked the supermarket manager for permission to carry out your survey!)
- the number, and kinds, of people who use fast food restaurants at various times of the day or night.

Market research surveys provide much more valuable data, but they take far more time and effort. (See the unit on market research in the companion volume in this series, *Markets*).

Rounding numbers

When you do your own local research, you will normally have insufficient time to collect masses of data, so big numbers will be no problem. However, when you are using published statistics, you will often come across big numbers such as

2,623,142

Large numbers, like these, are difficult to imagine and difficult to use in charts and graphs.

Any number can be rounded off to the degree of accuracy required. Let's take a simple example. Say you wanted to round off a decimal fraction to a whole number. For example, you might want to round off 12.1. Look at the figure following the units you are dealing in. If it is less than 5, you round down. If it is more than five, you round up.

For example, 12.*1* would be rounded *down* to 12, because the following figure *1* is less than five.

On the other hand, 12.*8* would be rounded *up* to 13, because *8* is more than five.

What happens when the following figure is five? In that case, the total is rounded to make the last figure an *even* number.

So 12.5 would be rounded *down* to 12, while 13.5 would be rounded up to 14.

The same principles apply, however big the number. For example, millions such as 2,623,*1*42 would be rounded down to 2,623 in thousands, because *1* is less than five.

In graphs and charts, numbers are usually rounded off to the nearest whole number or to one decimal point.

So, 2,623,142 would be rounded *down* to 2.6 million, because *2* is less than five.

Index

Abbreviations 15
Angles 6
Arithmetic mean 47
Arrays 47–8
Averages 47
Axes 8–9, 33, 57

Band curves 43–4
Baselines 40, 60–1
Borders 11–12, 29
Break-even charts 60–1
Broken scales 33

Cartograms 71–5
Combination charts 44–5
Comparisons 6, 9, 16, 20, 24
Component bar charts 23–6
Compound bar charts 19–22
Coordinates 33
Correlations 55–6
Curves 27, 28, 38–9, 45

Data 77
Demand curves 62
Dependent variables 30–1, 39
Design 9–12, 19, 28–9
Deviation bar charts 16–17
Direct variation 60
Distortion 31–2

Exploded segment 12

Fixed costs 60–1
Frequency 47, 51
Frequency distribution 48

Grid lines 28–9
Grouped frequency distribution 48
Guide lines 10

Histograms 47–51
Horizontal axes 8–9, 35
Horizontal bar charts 25

Independent variables 30–1
Index numbers 39–40

Labels 19, 23
Layer charts 43–4
Legends 19, 38–9
Line graphs 27–34, 36, 38–46, 53–4

Modal classes 48, 50
Multiple bar charts 19–22
Multiple line graphs 38–46

Negative values 16

Observation surveys 77–8

Percentage bar charts 24
Percentages 5–6, 10, 15
Perfect correlation 55–6
Pictograms 65–70, 72
Pie charts 4–7, 12
Population pyramids 52
Positive correlation 56
Positive values 16

Rounding 68, 78

Scales 27, 32–3
Scatter graphs 55–8
Segments 4–5, 12
Shading 11, 19, 24, 25
Single line graphs 27–34
Slices 4–5, 12
Software vi–vii, 5, 9, 10–11, 16, 24, 25, 63, 66
Sources 4, 76–8
Statistics 76–7
Stem-and-leaf diagrams 49–50
Step graphs 35–7
Straight line graphs 59–64
Strata charts 43–4
Supply and demand diagrams 63
Symbols 65–9

Tables 1
Tallies 47–8
Time 8–9, 15
Titles 4, 12

Values 28–9
Variable costs 61, 62
Variables 29–31, 39, 59
Vertical axis 9, 16, 33, 57
Vertical bar charts 1, 8–18, 27, 73
Vertical line charts 50–1, 53–4